GERMAN CAMOUFLAGED HELMETS OF THE SECOND WORLD WAR

Also by the Author

HELMETS OF THE FIRST WORLD WAR *(with Michael Haselgrove)*

GERMAN HELMETS OF THE SECOND WORLD WAR *(two volumes)*

GERMAN CAMOUFLAGED HELMETS OF THE SECOND WORLD WAR

Volume One: Painted and Textured Camouflage

Branislav Radovic

Schiffer Military History
Atglen, PA

Acknowledgements

It is with a deep sense of gratitude that the author would like to thank the men who own some of the world's best German World War II helmets. Without their help, these books would not have been possible – they gave their time generously and without condition.

A special thanks to John Burnett and David Asselin for their enthusiasm and drive, and to Chris Jones, Barry Smith and Bill Shea for being so generous with their collections. To my very good friends, Alan Beadle, Malcolm Fisher, Brian Maederer and Ellwood von Seibold, my thanks for all their support. I would like to thank Philippe Gillain, who gave so much of his valuable time at his splendid museum in Belgium, a must-see for any serious military collector. The fraternity of helmet collectors is a warm one indeed and I count myself fortunate to know them.

Many thanks go to my wife, Anna, who helped by turning Martian into English.

Lastly, thank you to Bob Biondi at Schiffer who has spent many hours in dark, hot and freezing conditions, sharing hotel rooms with a sick author and still coming out with his integrity intact ... and becoming a real friend.

The author would also like to explain that any errors in captions or credits are entirely his own and apologises to any contributor wrongly acknowledged or for any helmets incorrectly described.

In alphabetical order:

David Asselin
Alan Beadle
John Burnett
Martina Caspers (Bundesarchiv)
Alan Cant (Imperial War Museum)
Diana Condell (Imperial War Museum)
Malcolm Fisher
Perry Floyd
Philippe Gillain
John Hodgin
Chris Jones
Michael Kapitzke (Photography for David Asselin collection)
Brian Maederer
Tony Matassa
Bill Shea
Adam Smith
Barry Smith
Ty Smith
Matthew Tredwen
Ellwood von Seibold

These volumes are dedicated to those who struggle to keep all history alive and have not succumbed to current social pressure to remember only approved or sanitised subjects.

Book design by Robert Biondi and Branislav Radovic.
Photography by Robert Biondi.
Additional photography by Branislav Radovic and Michael Kapitzke.

Library of Congress Catalog Number: 2004101950.

Printed in China.
ISBN: 0-7643-2105-6

Published by Schiffer Publishing Ltd.
4880 Lower Valley Road
Atglen, PA 19310
Phone: (610) 593-1777
FAX: (610) 593-2002
E-mail: Info@schifferbooks.com.
Visit our web site at: www.schifferbooks.com
Please write for a free catalog.
This book may be purchased from the publisher.
Please include $3.95 postage.
Try your bookstore first.

In Europe, Schiffer books are distributed by:
Bushwood Books
6 Marksbury Ave.
Kew Gardens
Surrey TW9 4JF
England
Phone: 44 (0)20 8392-8585
FAX: 44 (0)20 8392-9876
E-mail: Bushwd@aol.com.
Free postage in the UK. Europe: air mail at cost.
Try your bookstore first.

Contents

Introduction

A BRIEF HISTORY OF MILITARY CAMOUFLAGE

Camouflage *n.* 1. the exploitation of natural surroundings, or artificial aids to conceal or disguise the presence of military units, etc. 2. (of fabric or clothing) having a design of irregular patches, in dull colours (such as browns and greens), as used in military camouflage. 3. the means by which animals escape the notice of predators. 4. a device or expedient devised to conceal or deceive. Origin 20C: from French, *camoufler*, from Italian, *camuffare*, to disguise. (As described by the *Collins Concise Dictionary*, 21st Century edition.)

Although the presence of camouflage within the animal kingdom has been with us for thousands, if not millions, of years, it was rather late in coming to humans. Notwithstanding the natural desire for any soldier, sailor or airman to make himself as invisible as possible in battle, it was not taken really seriously, nor officially recognised by the British, at least, until the second half of the 19th century, in places such as India, the Sudan and fighting the Boers in South Africa.

To take the British as an example, previosuly their uniform tunic was scarlet in colour, designed to disguise the obvious and demoralising effects of sabre, sword and small arms fire on the human body. The scarlet tunic was relegated to the parade ground and replaced, on active service, with a more subdued earth tone. The new colour was named khaki after the Urdu and Persian word for dust or earth – *khak*.

Khaki colour was also used for headgear, replacing the white versions of topee, or sun helmet, and, for the time being, that was the limit of the new thinking on personal camouflage.

First World War

By the First World War, nearly all the combatants were wearing, if not actual earth tones, much more subdued colours: khaki for the British and Belgians; field-grey for Germans and Austrians; and blue-grey for the French.

In terms of headgear, the German army produced clip-on cloth covers in field-grey for their leather and metal *pickelhaubes*, and also in darker grey and white versions for winter. These covers not only protected the rather delicate helmets, but, more importantly, masked the brightness of the steel and brass fittings.

The French army were the first to produce a, supposedly, ballistically tested steel helmet, the new Model 1915 "Adrian", after its designer, Quarter-master-general Adrian. At first it was painted a light blue-grey gloss, but later in a matte finish. Finding that these early helmets tended to reflect too much light, particularly at night, they also produced a cloth cover, usually of two halves sewn together, in various shades of blue and light tan and attached to the helmet by a drawstring around the rim.

The Italian army initially adopted the French helmet design and eventually produced its own version, the Model 1916, which was painted olive green. They too produced cloth covers in the same style as their Gallic allies, but unlike the French, the Italians added insignia by stencilling, embroidery and sewn-on cloth badges.

The British and German armies were quick to follow the new trend and produced their own designs of steel helmet. In a remarkable, if not unique, flash of foresight, the earliest British helmet of 1915, named "Brodie" after its designer John Leopold Brodie, was initially produced in camouflage colours of orange and green, but the later production models were changed to apple green and then, lastly, to dark textured khaki. Perhaps the camouflage paint scheme was thought too radical, or expensive, to produce in the vast quantities that would be required. Cloth covers, made of canvas or hessian (burlap) were produced from one, two or more pieces, in various shades of tan and these too were attached to the helmets by a drawstring around the rim.

The German army's steel helmet made its first appearance in early 1916, around Verdun. It was painted smooth field-grey, which like the French,

Italian and British versions became rather shiny with wear or when wet. In order to address this problem, cloth covers of tan, field-grey and white, for winter, were produced. These often had leather reinforcing to fit around the ventilation holes that supported the armoured browplate, and again, were secured to the helmet by drawstring or metal hooks that clipped around the rim.

None of these combatant nations were able, or willing, to produce enough of these camouflage covers to equip all their troops and so, inevitably the troops produced their own versions from materials at hand, often hessian or canvas, and in a myriad of designs and colours.

It was apparently the German High Command who were the first officially to sanction the application of camouflage paint schemes for helmets. In July 1918, it was ordered that if German troops wished to camouflage their steel helmets, they were to use no more than four colours, reflecting the surrounding seasonal tones, and to outline each colour with a finger's width of black, thus helping to break up the shape of the helmet.

Germany's Austrian allies produced their own steel helmet in 1916, the Berndorfer, but this design was superseded a year later by a copy of the German Model 1916, which was painted khaki. Cloth covers were not as much in use as in the German armed forces, but recorded examples are very similar to their German counterparts. There are also examples of Austrian-made steel helmets carrying the German painted camouflage scheme, described above.

Curiously, many surviving examples of Brodie-style American helmets are painted in various colours with the black outline, mimicking the German practice. Obviously, many US troops were greatly impressed with the German patterns, but it is not clear whether they actually wore these copied designs into battle. It seems more likely that they were painted as souvenirs to be brought home.

Inter-War Period

The immediate post-war period saw few real advances in the practice of camouflaging helmets, although some attempts were being made.

During the late 1930s, Dr Brandt of the SS-*Verfugungstruppen*, was experimenting with several designs of camouflage material that emulated various terrain and foliage. This coordinated camouflage would be used by the new politically-motivated arm of service that would, in March 1940, be re-named the Waffen-SS. These patterned designs were used on a sprung clip-on cloth cover, which was reversible from predominantly green for spring, to rust/brown for autumn. The covers were later produced with the addition of cloth loops to facilitate the attachment of local foliage to break up the helmet outline. These new SS designs were by far the most radical concepts in camouflage techniques to date and can even be thought of as the precursors of current usage.

The German army adopted a camouflage pattern as early as 1932. It is what collectors have come to call the "splinter" pattern – a design of jagged blocks of green and brown over a background of light green with thin strips of brown. The fabric was used for the poncho/tent quarter (*zeltbahn*) and seems not to have been used for camouflaging helmets until the mid-war period.

The inter-war period also saw the French military authorities continue to produce various cloth helmet covers in the same style as the Great War. Examples can be found made from a fairly close weave cotton and are usually made of two halves, with or without sewn tucks across the seam, in light khaki or white (presumably for mountain troops). They upgraded the design of the Adrian helmet with thicker steel, fewer pieces and a better liner system. Their uniforms were changed from blue-grey to khaki, and their helmets were painted to match.

The military forces of Imperial Japan were issued with a cloth helmet cover, as can be seen in photographs of Japanese troops in China in the mid-1930s. They were made from several panels of light khaki cloth, were padded to insulate against the heat, and attached to the helmet by drawstring. They often carried the insignia of the army (a yellow star) or navy (an anchor) embroidered on a small patch and sewn onto the front. Later photographs show these covers sometimes being worn with string netting over the top for attaching foliage, so important for fighting in the forests and jungles of the Far East.

Elsewhere, amongst the other soon-to-become combatant nations, this truly co-ordinated approach to camouflage and particularly the camouflaging of the steel helmet was not yet being adopted.

Second World War

By 1940/41, we see photographs of British troops in the Low Countries and Norway, wearing hessian covers, apparently a regulation type with loops for foliage, also many that seem to have been field-made. They were attached to the helmets by a drawstring and it is fairly clear these covers are direct descendants of the covers worn by troops in the 1914-18 war.

Several patterns of drawstring netting covers were also produced and issued on a large scale, particularly later in the war in the Far East and during the Liberation of Europe. British paratroops also wore two-colour netting covers and, from 1944, it became common practice throughout the land and airborne forces to add strips of hessian to the netting to disrupt the outline.

Early wartime photographs can be found of British helmets being worn with bands of wire attached. At least one clear example in the photographic archives of the Imperial War Museum in London, shows Medical Corps personnel attached to an armoured division, wearing this wire configuration over camouflage painted (black and green) helmets, during training. Another familiar style of camouflaged British helmet can be found in photographs of troops in Malta during the siege. These helmets were painted with patches of dark brown/khaki over a sand base colour. This was probably not a regulation pattern, but one becoming rather popular and widespread through practice and example, creating its own *esprit de corps*.

The United States of America appears not to have made very much use of camouflage, particularly for helmets. This probably reflects the isolationist, and essentially non-military, attitude of the American people at that time. Later in the war, however, a purpose-made regulation camouflage cover was produced for the Marine Corps. It was reversible from predominantly greens one side to browns on the other and was subsequently developed with small buttonhole-sized slits for foliage. Various patterns of string netting covers were also produced and attached by drawstring or by passing the bottom of the netting between the liner and the helmet shell.

It could be argued that the Italians were the first to produce a truly modern helmet shape, the Model 1933. The Futurist movement was sweeping the art world in the 1920s and 1930s and efforts to modernise their armed forces and its equipment reflected this. One only has to examine their medals and insignia to see the new streamlined approach to design emerging. A non-reversible light tan cloth cover was produced with a drawstring that carried an ink stamp inside of the king (VE) Vittorio Emanuele. It was manufactured in several sizes and was made from two halves sewn together. Later, during the war, other variations were created, some with a distinctive multi-coloured pattern and also some attached to the helmet by small hooks. Italian paratroops had their own covers, in the patterned material and with loops for foliage.

Having set the scene, let us focus on the German armed forces. At the start of the war in 1939, in Poland, it was fairly common practice to use the rubber tyre inner tubes to hold foliage around the circumference of the new model 1935 helmet. Needless to say, very few examples, if any, now exist with the original rubber attached. Another practice was to use a leather belt or the helmet carrying harness to hold camouflage cloth to the helmet. Regulation army camouflage material was not readily available at that time and so the strangest scraps of patterned cloth were being used. These were probably taken from the furnishings of homes deserted by their owners, or dress material taken from shop windows.

The earliest examples of true regulation camouflage covers are found in photographs of Waffen-SS troops in Holland in 1940. These are the early pattern, previously mentioned, before the addition of foliage loops, and are often pictured being worn by troops with the matching camouflage smocks. A matching, reversible spring-to-autumn face veil of camouflage printed strands was also produced to hide the face. The strands were thrown back over the top of the helmet when not in use.

Around 1941, the German army extended their use of the "splinter" camouflage material to helmet covers, usually of the drawstring type although examples of hook and sprung clips do exist, both with and without foliage loops. They were produced in both non-reversible and reversible (to white for winter fighting) formats. However, as in the First World War, field-made covers were common, most particularly in North Africa. If helmets were not painted in the tan vehicle colour, men would fashion covers from canvas or sandbag material.

The Lufwaffe paratroop (fallschirm) service developed its own cloth cover of a grey-green material for their Model 1938 steel helmet. It had a brown cloth band sewn around the circumference with a crossed loop on the top and was secured to the helmet by six hooks. Its first use seems to have been in autumn 1940. A later version used the same construction but adopted the army-style "splinter" pattern fabric and was first used in Crete in May 1941. Later still, a "blurred splinter/tan-water" patterned material was introduced for clothing and was used for field-made covers. Subsequent versions were also developed with a drawstring attachment instead of hooks.

Netting covers were also produced for the army and Luftwaffe, the most common being an open weave design of knotted string which could be attached to the helmet by tucking the netting down between the liner band and the shell, or by using hooks to clip under the helmet rim. A drawstring threaded through the netting around the circumference helped to keep the whole thing tight onto the helmet shell. An extended piece of netting at the front of the helmet was used as a face veil and could be held back over the helmet by two small hooks. Other types of netting existed, some attached by hooks instead of drawstring. Great use was made of captured helmet netting, particularly in the fighting around the Anzio, Cassino and Normandy fronts.

The most individual, and perhaps the most interesting, form of helmet camouflage is the owner-painted type. As we have already seen, the practice of painting camouflage colours directly onto the helmet surface began, in earnest, during the First World War. Although it became fairly common to all armies of the Second World War, it was particularly prevalent in the German armed forces.

The most common types were: the simple expedient of dulling down the factory finish with matt paint; the use of sand-coloured vehicle paint in North Africa, sub-tropical Italy or Soviet Union; water-soluble whitewash for winter campaigning, which could be scrubbed off in spring or spirit-based white paint which could be over-painted; multi-colour schemes of tan, brown and green to blend in with the local flora; and combinations of contrasting shades, sometimes of the most unlikely colours, to create a disruptive or "dazzle" effect. In addition, various elements, such as sand, grit or wood chips, were added to the paint to alter the texture and reflective qualities. This is often referred to as "*zimmerit*", after the anti-magnetic mine paste applied to armoured vehicles during the second half of the war.

The protective steel helmet was one of the most vital and personal pieces of equipment a soldier possessed. As I hope this book will show, all manners of systems and styles of camouflage coverings were used – bands of string, wire and strips of steel, with and without cloth covering – making the variations practically endless. So too are the painted helmets with every conceivable pattern, colouring and texture, reflecting the skills of the individuals who produced them. Some are of the greatest technical quality and intricacy; some are bold, while others are subtle. Many are very crude and some verge on the artistic.

On one level, camouflage helmets are wonderful and desirable in their own right as military collectibles. On another, deeper, level, each one gives us an insight into the soldier's awareness of the world around him, how seriously he took the art of concealment and perhaps, even, a little of his character, which he reveals, albeit inadvertently, by his choice of colours, patterns and materials.

Chapter One: Painted Camouflage

M35, originally painted the pre-war apple green. It was later over-painted with brushstrokes of matte green, tan and brown, leaving only the Wehrmacht eagle insignia visible. (Collection of Perry Floyd)

M35, the original colour has been completely re-painted with brushstrokes of thick matte green and tan and stripes of brown. (Collection of Chris Jones)

M35, the factory colour was completely over-painted in white for winter camouflage, then repainted dark green, either for seasonal change or theatre of operations. This helmet shows traces of its original SS runes insignia under the paint. (Collection of Ty Smith)

M35, which seems to have been originally painted dark field-grey, then completely over-painted with tan and sprayed with dark green. (Collection of John Burnett)

M35, originally painted smooth Wehrmacht mid-green, then over-sprayed with stripes of tan and brown. (Collection of Brian Maederer - Military Collectibles)

M35, originally painted Luftwaffe blue-grey, then completely over-painted in a dark, almost black, colour with patches of tan. The twin insignia of the Luftwaffe have been hand-painted over the top. (Collection of Brian Maederer - Military Collectibles)

M35, originally painted the pre-war apple green, then over-sprayed with matte tan and stripes of dark green. (Collection of author)

M35, originally painted dark field-grey, completely over-painted with a thick coat of mid-green and patches of red/brown. (Collection of Tony Matassa)

M35, originally painted in the darker mid-war Luftwaffe blue-grey, then over-painted with patches of green and brown. The Luftwaffe second pattern eagle insignia is just visible. (Collection of Brian Maederer - Military Collectibles)

M35, one half completely over-painted in dark green, the other in red brown, with patches of tan applied over the whole. (Collection of Adam Smith)

M35, originally painted in the pre-war apple green, then completely over-painted with matte dark green. The twin insignia of the Wehrmacht are visible. The paint is worn away around the skull, suggesting wear from a wire or rubber inner tube. This badly damaged helmet belonged to *Feldwebel* Zander who survived the damage and was awarded the German Cross in Gold and Close Combat Badge in 1945, and was promoted to *Leutnant*. (Collection of the author)

Zander

M35, originally painted smooth dark blue-grey then over-painted in tan and sprayed with patches of dark green. The twin insignia of the Luftwaffe with second pattern eagle have been left visible. (Collection of Malcolm Fisher - Regimentals)

M35, originally painted dark blue-grey, then over-painted with patches of tan. It carries the second pattern Luftwaffe eagle insignia and has a period applied U.S. webbing chinstrap. (Collection of David Asselin)

M35, originally painted smooth Luftwaffe blue-grey, then completely over-painted in tan and sprayed patches of dark green. The national shield has been over-painted while the first pattern eagle had been left visible. (Collection of Barry Smith)

Gustav Reckmeier

Veteran Corporal Jack L, Kaplan, 347th General Service Regiment, 3rd U.S. Army, holding the helmet he took as a souvenir. It is named to *Obergefreiter* Gustav Reckmeier 3. *Batterie Schwere Gemischte Flakabteilung* 153. Based in Cherbourg, Normandy, he was killed in action on June 24, 1944 and is buried in the German War Cemetery in Orglandes, Normandy.)

M35, originally painted Wehrmacht green, initially over-painted with textured dark green then completely over-painted in tan. (Collection of John Burnett)

M35, the original dark grey-green has been completely over-painted in tan. (Collection of John Burnett)

M35, originally painted Luftwaffe blue-grey then completely over-painted in white. This helmet was picked up outside St. Vith, Belgium by veteran Robert A. McKinsey in 1946. (Collection of David Asselin)

M35, overpainted dark grey-green and with khaki patches, The Kriegsmarine eagle insignia has been painted around. (Collection of John Hodgin)

M35, sand/grit texture oversprayed dark grey-green, tan and brown. The eagle insignia is probably army and has been oversprayed but remains visible. (Collection of John Hodgin)

M40, oversprayed green, tan and brown. The Kriegsmarine eagle has been oversprayed but remains visible (Collection of John Hodgin)

M40, oversprayed tan with patches of a dull olive green. Again, the eagle insignia has been oversprayed and is barely visible. (Collection of John Hodgin)

M40, originally painted in dark green (Wehrmacht?), over-painted in tan with splashes of green. (Collection of David Asselin)

M40, originally painted dark green over-painted in white with sprayed spots of bright green. It carries the hand-painted image of an owl at the front. (Collection of John Burnett)

M40, originally dark green, over-painted in tan and sprayed with dark green stripes. (Collection of John Burnett)

M40, with its original dark green matte finish over-painted in stripes of tan. It carries the second pattern Luftwaffe eagle insignia. (Collection of John Burnett)

M40, originally dark blue-grey, completely over-painted with tan. The first pattern Luftwaffe eagle insignia has been left visible. (Collection of Chris Jones)

M40, originally dark grey-green completely over-painted with white. The second pattern SS rune insignia has been left visible. The original owner was a *Kriegsberichter* (war correspondent) and is marked inside as SS *Kriegsberichter Frei.Leg.Norwegen.* (Collection of Chris Jones)

M40, the original dark grey-green has been completely over-painted in tan, with the first pattern SS rune insignia left visible. (Collection of Alan Beadle Militaria)

M40, the original dark grey-green has been over-sprayed with two tones of green. The SS rune insignia is barely recognisable through the paint. (Collection of Chris Jones)

M40, originally dark grey-green, over-painted in textured tan finish. (Collection of Chris Jones)

M40, over-painted in a matte chocolate brown with green patches over the top. (Collection of John Burnett)

M40, originally blue-grey then over-painted in green with brown patches. The second pattern Luftwaffe eagle insignia has been left visible. Note the use of a non-regulation chinstrap from a Polish M31 helmet. (Collection of Phillippe Gillain - December 44 Museum, La Gleize, Ardennes, Belgium)

M40, originally dark grey-green, over-painted in tan and sprayed over with patches of green and brown. (Collection of Brian Maederer - Military Collectibles)

Dehne

M40, originally dark grey-green, over-painted in tan and sprayed with light spots of red/brown. The Wehrmacht eagle insignia, though extremely worn, has been left visible. (Collection of Brian Maederer - Military Collectibles)

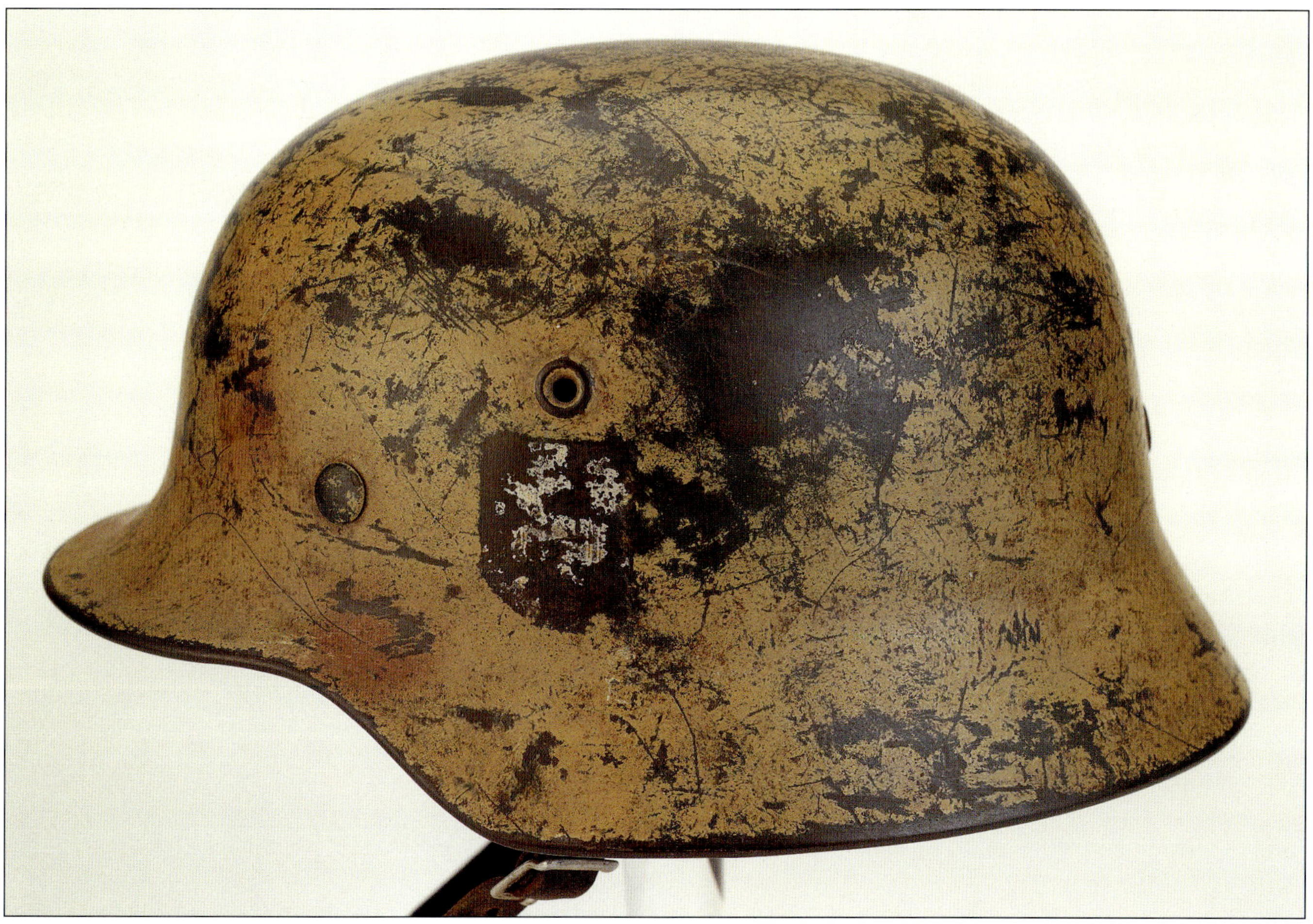

■ M40, originally mid-to-late war dark matte grey-green, over-painted with patches of tan and emerald green. The second pattern Luftwaffe eagle insignia appears to have been placed over the tan layer. (Collection of Brian Maederer - Military Collectibles)

M40, originally mid-to-late war dark matte grey-green, completely over-painted with strips of matte grey, light and mid greens. The Wehrmacht eagle insignia has been left visible. (Collection of Brian Maederer - Military Collectibles)

M40, originally mid-to-late war dark matte grey-green, over-painted dark matte green and patches of white. The second pattern Luftwaffe eagle insignia is partially visible under a patch of white. (Collection of David Asselin)

M40, originally mid-to-late war dark matte grey-green, over-painted white with stripes of bright green. The Wehrmacht eagle insignia has been left visible. (Collection of the author)

M40, originally mid-to-late war matte dark grey-green, over-painted in tan with sprayed patches of green. The Wehrmacht eagle insignia has been completely over-painted. This helmet was brought back to England by British veteran Corporal Don Clough of the Royal Signals Corps. It was picked up in Bruhl, near Cologne, while he was attached to the U.S. Army in the autumn of 1944. (Collection of the author)

M40, originally painted in mid-to-late war matte dark grey-green over-painted in tan then sprayed with patches of green and brown.

M40, originally painted in mid-to-late war matte blue-grey, over-painted in tan with the Luftwaffe second pattern eagle insignia left visible. (Collection of Barry Smith)

Battle-damaged M40, originally painted in mid-to-late war matte dark grey-green, completely over-painted in tan with a Wehrmacht eagle insignia left visible. (Collection of Matthew Tredwen)

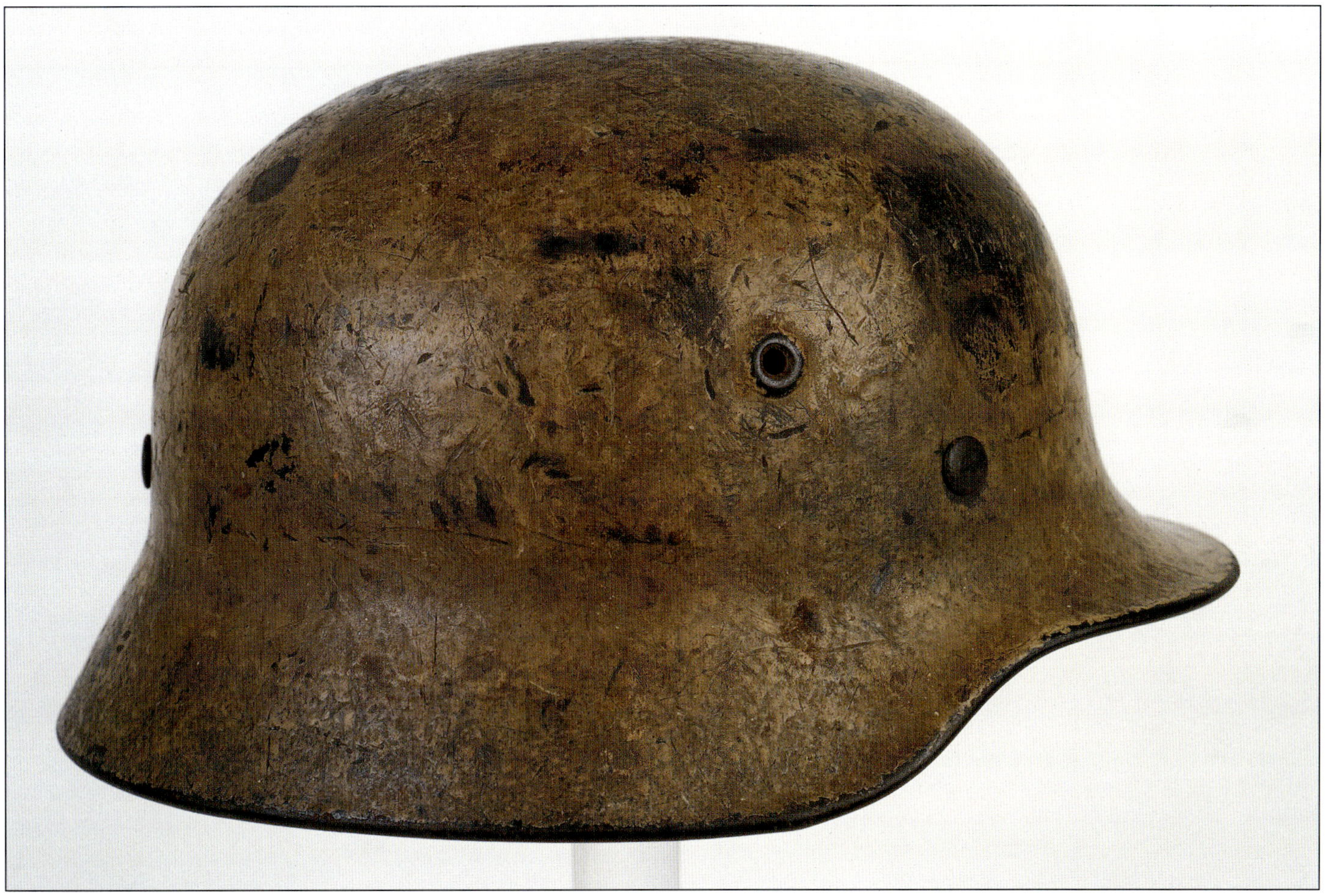

M40, originally painted in mid-to-late war matte dark grey-green, over-painted in tan and sprayed with patches of green. The outline of the Wehrmacht eagle insignia is recognisable under the paint. (Collection of Imperial War Museum, London)

M40, originally painted in mid-to-late war blue-grey, over-painted with a mixture of tan paint and sand. The national shield and Luftwaffe second pattern eagle are visible under the paint (Collection of John Burnett)

M40, originally painted in mid-to-late war blue-grey, painted in patches of pinkish tan and light green. (Collection of John Burnett)

M40, originally painted in mid-to-late war blue-grey, completely over-painted in mustard tan. The Luftwaffe second pattern eagle insignia has been picked out in white. (Collection of John Burnett)

M40, originally painted in mid-to-late war matte dark grey-green, completely over-painted in tan. The Wehrmacht eagle insignia is quite recognisable under the paint. (Collection of Perry Floyd)

M40, originally painted in mid-to-late war matte dark grey-green, with traces of the whitewash camouflage remaining. The state shield and SS rune insignia have been left visible. (Collection of Alan Beadle Militaria)

M40, originally painted in mid-to-late war blue-grey, completely over-painted in white. The outline of the Luftwaffe second pattern eagle insignia is visible under the paint. (Collection of John Burnett)

M40, completely over-painted in white. (Collection of John Burnett)

M40, originally painted in mid-to-late war blue-grey, completely over-painted in white, with a large red cross painted on the top. The outline of an insignia exists on the left side. (Collection of Chris Jones)

M40, originally painted in mid-to-late war grey-green, completely over-painted in white with a small red cross painted on the top. The Wehrmacht eagle insignia is barely recognisable under the paint. (Collection of John Burnett)

M40, originally painted in mid-to-late war grey-green, completely over-painted in tan and sprayed in patches of green and brown. It has "FRANCE JULY 14 1944" painted inside the rim. (Collection of David Asselin)

■ M42, originally painted in mid-to-late war blue-grey, completely over-painted in tan with stripes of dark green. An inverted triangle of thinner tan paint has been painted around the Luftwaffe second pattern eagle. (Collection of John Burnett)

M42, originally painted in mid-to-late war blue-grey, completely over-painted in tan, except for an inverted triangle around the Luftwaffe second pattern eagle insignia and then over-sprayed with patches of dark brown. (Collection of the author)

M42, originally painted in mid-to-late war grey-green, completely over-painted in brown with one patch of matte green sprayed onto the right side. The second pattern SS rune insignia has been left visible. (Collection of Ty Smith)

M42, originally painted in mid-to-late war grey-green, over-painted in dark brown with thick strips of dark green. (Collection of David Asselin)

M42, originally painted in mid-to-late war grey-green, over-painted in large geometric shapes in white, light brown and dark green. (Collection of the author)

M42, originally painted in dark grey-green. It has been completely over-painted in tan and sprayed with patches of brown over a wire cage, now removed. (Collection of Brian Maederer - Military Collectibles)

M42, originally painted in dark grey-green, over-painted with patches of brown and light and dark green. (Collection of John Burnett)

M42, originally painted in dark grey-green, completely over-painted in tan and patches of matte light green. The Wehrmacht eagle insignia has been left visible. (Collection of David Asselin)

M42, originally painted in dark grey-green, completely over-painted in white. The Wehrmacht eagle insignia is visible through the paint. (Collection of David Asselin)

M42, originally painted in dark grey-green, completely over-painted in white with a large red cross over the top. (Collection of John Burnett)

M42, originally painted in dark grey-green, completely over-painted in white with a small red cross over the top. A British veteran picked up this helmet in the Ardennes area. (Collection of Brian Maederer - Military Collectibles)

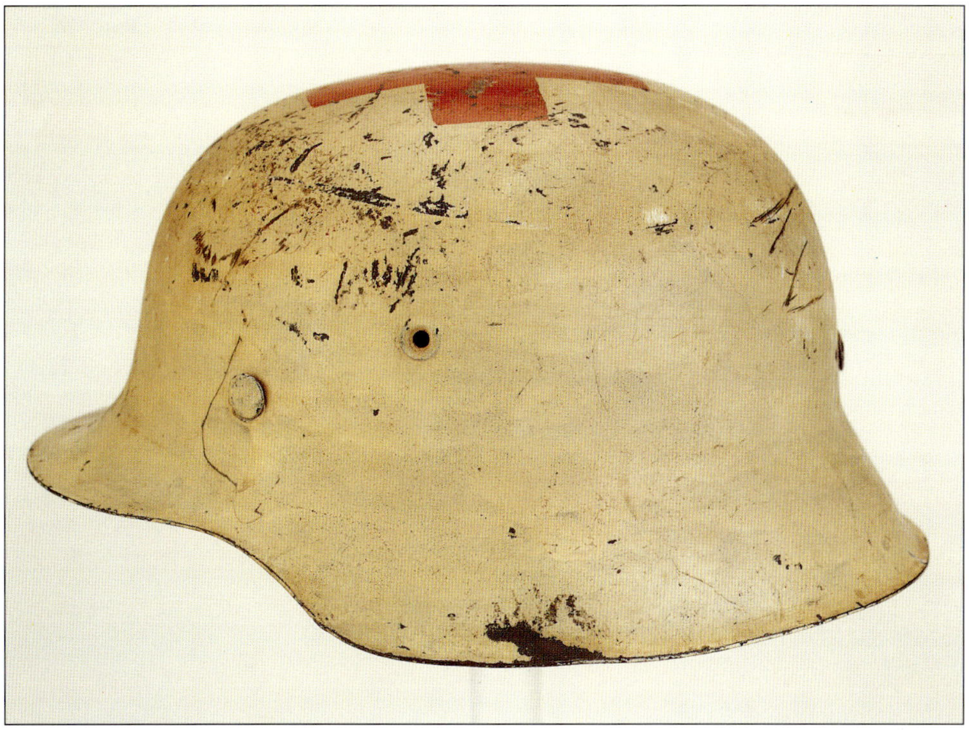

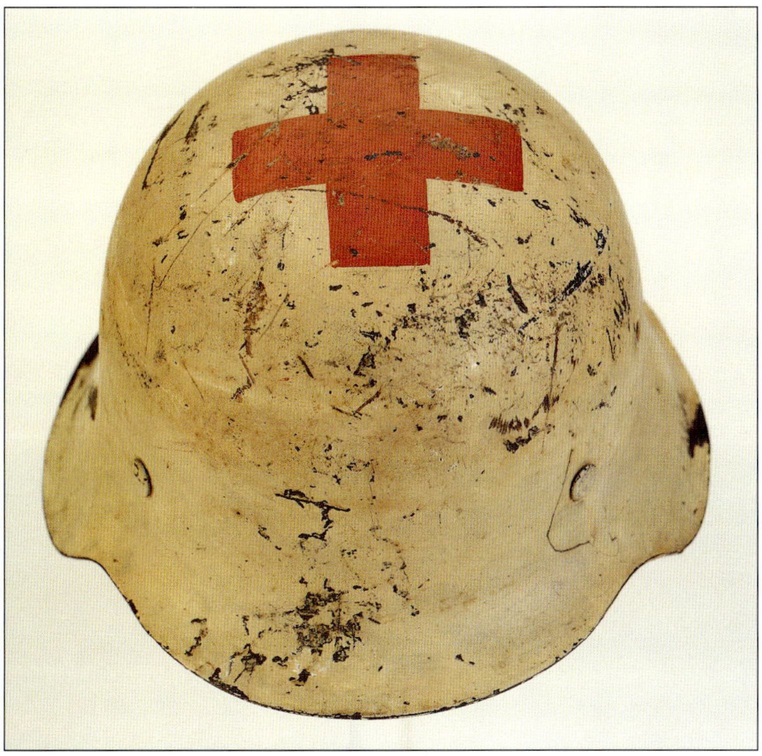

M38 shell, originally painted pre-war apple green, partially over-painted in tan. The national shield and Luftwaffe second pattern eagle have been left visible. (Collection of the author)

M38, originally painted mid-to-late war grey-green, completely over-painted in tan with patches of mid-green. (Collection of Chris Jones)

M38, originally painted dark grey-green, completely over-painted in mustard tan. The Luftwaffe second pattern eagle insignia is barely visible through the paint. (Collection of Chris Jones)

M38, originally painted dark grey-green, over-painted with patches of tan. Its Luftwaffe second pattern eagle insignia is the subdued type. The liner/ventilation bolts are the late-war aluminium solid slotted type. (Collection of Chris Jones)

M38, originally painted dark grey-green, completely over-painted in mid-green and patches of tan over a sand/grit base. The Luftwaffe second pattern eagle insignia is easily visible through the paint. (Collection of Chris Jones)

M38, originally painted dark grey-green, completely over-painted in tan and sprayed with patches of mid-green. (Collection of the author)

M38, originally painted dark grey-green, completely over-sprayed with patches of tan and mid-green over a sand/grit base. The well-worn Luftwaffe second pattern eagle insignia is clearly visible through the paint. (Collection of John Burnett)

M38, originally painted dark grey-green, completely over-painted in white. The well-worn Luftwaffe second pattern eagle insignia is visible through the paint. This helmet was left in La Gleize in the Ardennes, Belgium, and picked up after the battle for the town. (Collection of Phillippe Gillain - December 44 Museum, La Gleize, Ardennes, Belgium)

M38, originally painted dark grey-green, completely over-painted in tan. The Luftwaffe second pattern eagle insignia is clearly visible through the paint. Note the field-made replacement bolt at the rear. (Collection of Imperial War Museum, London)

M38, originally painted dark grey-green, completely over-painted in mustard tan and patches of light green. (Collection of John Burnett)

M38, originally painted dark grey-green, completely over-painted in tan and sprayed with patches of green over a sand/grit base. (Collection of John Burnett)

M38, originally painted pre-war apple green, over-sprayed with patches of tan and green. The national shield and Luftwaffe second pattern eagle insignia are visible through the paint. (Collection of John Hodgin)

M38, originally painted dark grey-green, completely over-painted in tan and sprayed with patches of brown and green. (Collection of John Hodgin)

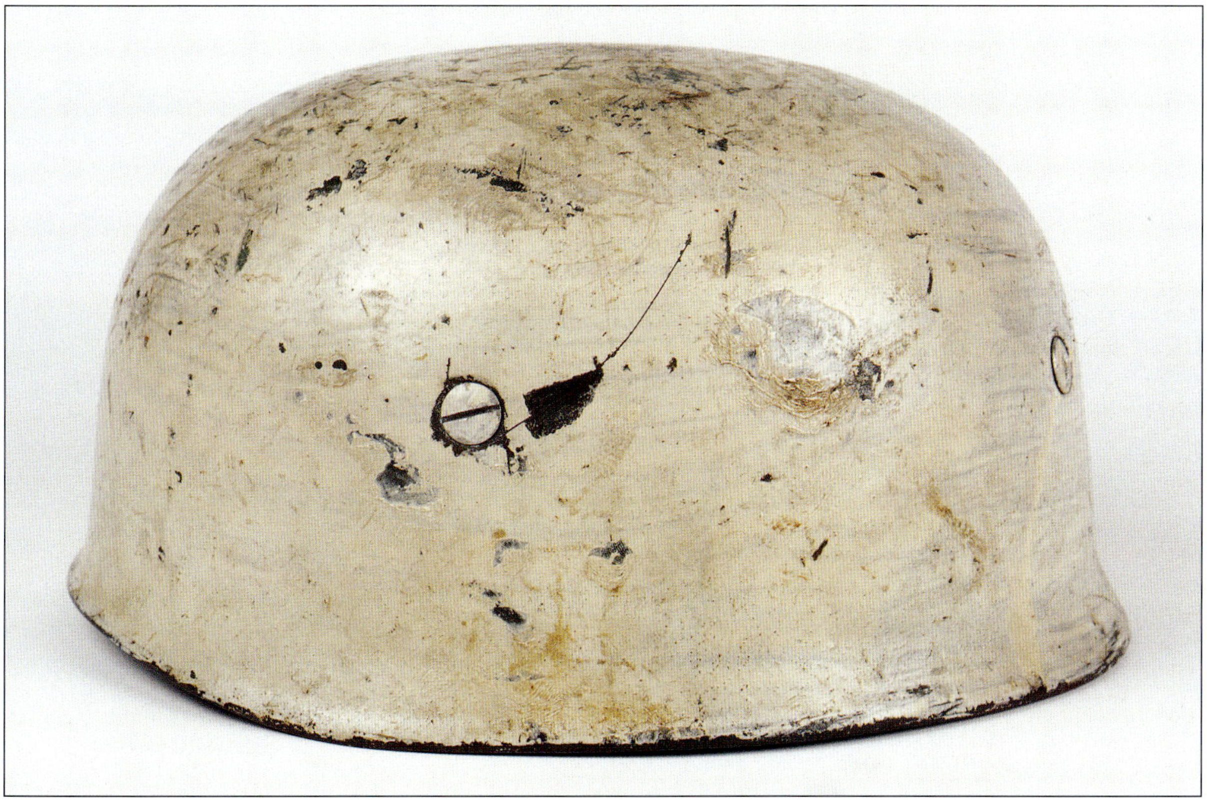

M38, originally painted dark grey-green, completely over-painted in white. The liner/ventilation bolts are the late-war aluminium solid slotted type. (Collection of John Hodgin)

M38, originally painted dark grey-green, over-painted in stripes and patches of tan, red-brown and green. The Luftwaffe second pattern eagle insignia has been left visible. (Collection of John Hodgin)

M38, originally painted dark grey-green, completely over-painted in khaki with patches of red-brown and flecks of tan. The Luftwaffe second pattern eagle insignia has been left visible. (Collection of John Hodgin)

M38, originally painted dark grey-green, over-painted with light green and red-brown. The Luftwaffe second pattern eagle insignia is clearly visible through the paint. (Collection of John Hodgin)

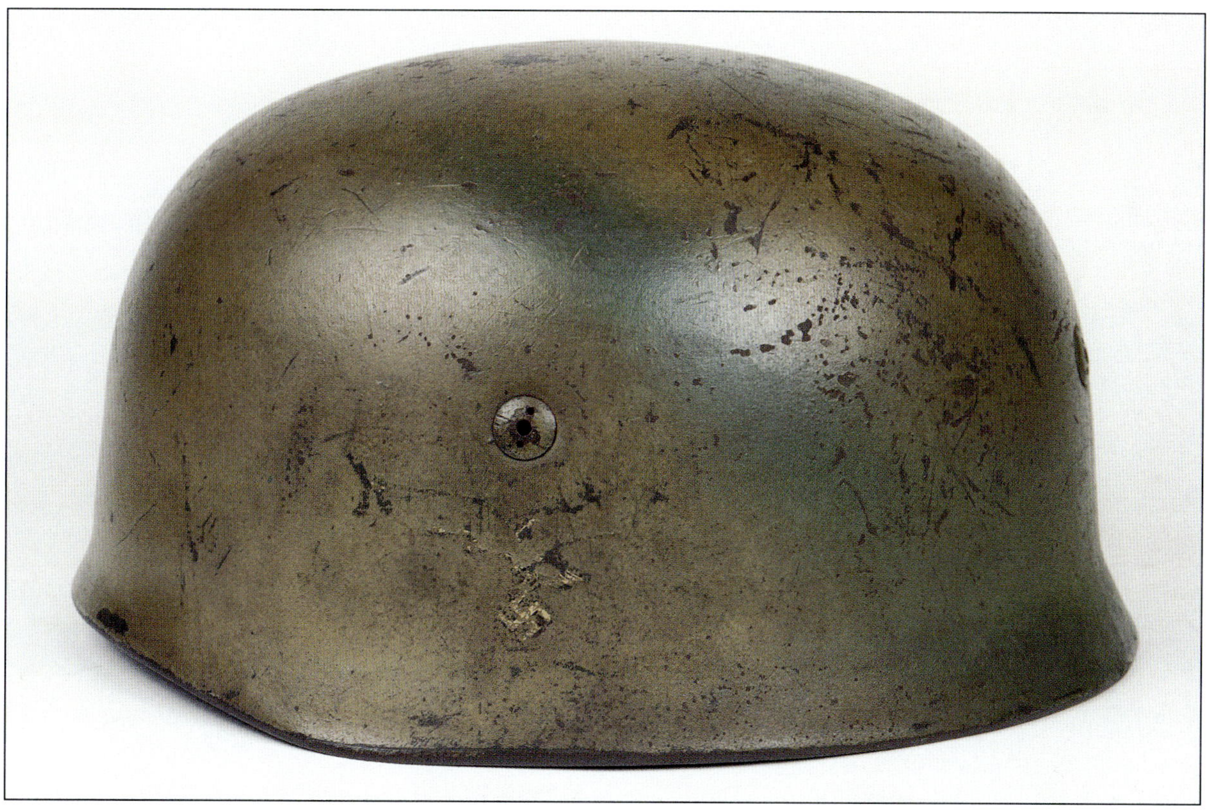

M38, originally painted dark grey-green, completely over-painted in tan and sprayed with patches of light green. The Luftwaffe second pattern eagle insignia is visible under the paint. (Collection of John Hodgin)

Period Photos

A motorcyclist wearing an M40 helmet with a typical example of the three-coloured "Normandy" style camouflage paint scheme. *(BA)*

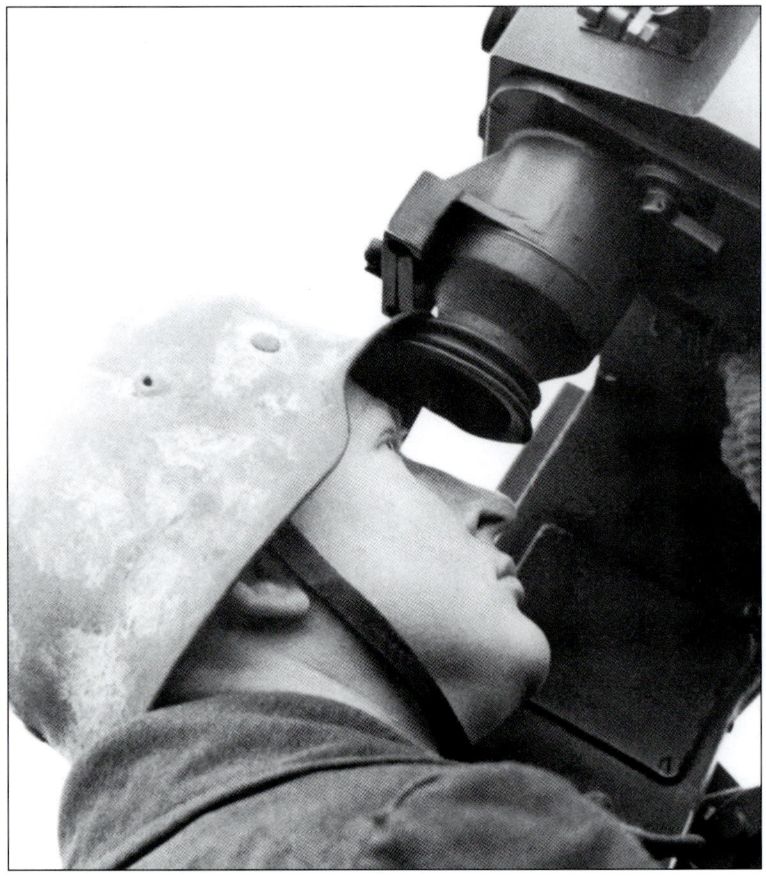

Top left: Another example of the "Normandy" camouflage scheme. *(BA)*

Top right: This anti-aircraft gunner's helmet appears to have been painted with light sand colour patches over a darker base. *(BA)*

Bottom: Young troops from the 12th SS *Hitlerjugend* division attacking British and Canadian troops on 7th/8th June 1944, near the Abbaye d'Ardenne in Normandy. Two of them are wearing sprayed three-tone camouflage helmets, while the third has the issue late pattern Waffen-SS reversible three-clip helmet cover. *(Author)*

Members of the Hermann Göring Division wearing Waffen-SS camouflage smocks, in conversation with an Italian officer. The man on the left has sprayed his M35 with a light tan base colour and then sprayed a darker (green?) over foliage to achieve an extremely creative effect. *(BA)*

■ Top left: Taken in September 1944, a group of Kriegsmarine men, crewing an anti-aircraft gun. They wear an interesting collection of helmets including an M16, an M42 and two M35s. All helmets appear to be painted "battleship" grey. *(BA)*

■ Top right: Photographed in August 1944, this NCO Panzer crew member is wearing an over-painted M42, leaving the Army eagle insignia only partially visible. *(Author)*

■ Bottom left: *Generalfeldmarschall* Fedor von Bock awards troops in the field in November 1941. Their M35s are well used and over-painted, again, obscuring the insignia. *(Author)*

■ Bottom right: This veteran wears a light tan painted M40 helmet. *(BA)*

Top: These men, manning a defensive position on the Russian front, have all painted their helmets solid white. *(Author)*

Bottom: More white-painted camouflage, but these men have left areas unpainted to break up the helmet's outlines. *(Author)*

Top: Taken in May 1944, southwest of Narwa in the USSR, one man in this group has painted his helmet with white stripes to break up the outline. *(Author)*

Bottom: These men proudly sport their newly issued helmets, M35s or M40s. It is interesting is that some of the helmets have received a matte painted layer over the older pre-war smooth finish. This illustrates the very first step in camouflaging helmets to suit combat conditions. *(Author)*

A clear example of the "Normandy" paint scheme on an M40. *(BA)*

Taken at the same time and place as the previous photograph, this man is wearing the same camouflage scheme, this time on an M42. A veteran of the first winter in the USSR, he could be wearing a *Krim* campaign shield on his left arm. *(BA)*

A group of *Fallschirmjäger* in Italy in the Anzio-Nettuno invasion front. Here we see a collection of painted and mud-spattered helmets. The man in the centre wears the regulation clip-on splinter pattern camouflage cover. *(BA)*

Top: These mid-to-late-war *Fallschirmjäger* have all had their M38 helmets sprayed with a light base colour and over-sprayed with what are probably green stripes. *(BA)*

Bottom: Another photograph taken during training, their helmets resting on parachute packs. As you will see, not all have had the same treatment – note the early twin insignia example. *(BA)*

Top: Another view of the same helmets. *(BA)*

Bottom: Taken at the same time and place as the previous three photographs, another group of *Fallschirmjäger* wear the plain regulation matte finish with eagle insignia. *(BA)*

A Fallschirmjäger, possibly in Rome, is wearing his M38 camouflaged with meandering lines of darker paint over a light base. (Note the missing liner bolt, a recurring problem.) *(BA)*

Top left: This *Fallschirmjäger* has over-painted his M38 either with an overall light colour with a little of the original dark base showing through, or an all over light tan base with a few darker patches over the top. *(BA)*

Top right: Probably produced in the same way, this man has achieved a more patchy effect. (Note his comrade behind, wearing the regulation drawstring splinter cloth cover.) *(BA)*

Bottom left: These two *Fallschirmjäger* in Tunisia have painted their helmets in a tan finish. The man on the right has added dark wavy lines, probably green. *(Author)*

Bottom right: This photograph, taken at Cassino in May 1944, illustrates the wartime grey matte paint that replaced the pre-war smooth green finish of M38s. They are wearing their parachute harnesses over their jump smocks while being presented to Herman Göring. *(Author)*

A close-up of this veteran Fallschirmjäger, also seen in an earlier photograph, shows his multi-coloured paint finish M38. His comrade is wearing a regulation clip-on cover, possibly the early grey-green version. *(BA)*

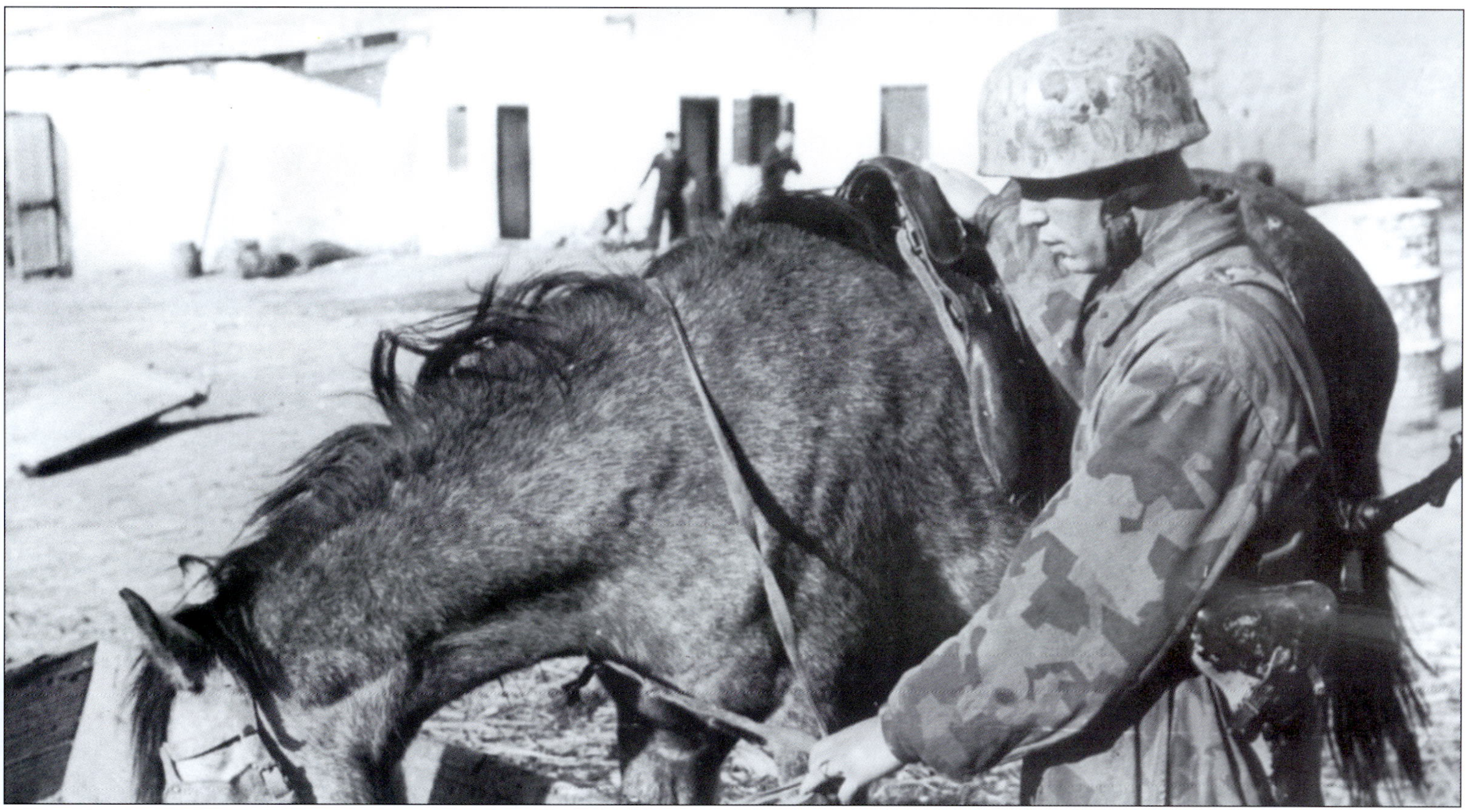

Top left: These Fallschirmjäger, shown in St. Peter's Square, Rome, are all sporting a camouflage pattern which seems to have been quite a favourite of men fighting around Anzio. It is a light tan base with uneven loops of dark green painted over the top. *(Author)*

Top right: This man in tropical uniform has completely over-painted his helmet in tan. *(BA)*

Bottom: A *Fallschirmjäger* feeds his horse in Tunisia in March 1943. His helmet is entirely covered in a light tan base colour with splotches of darker paint over the top. *(Author)*

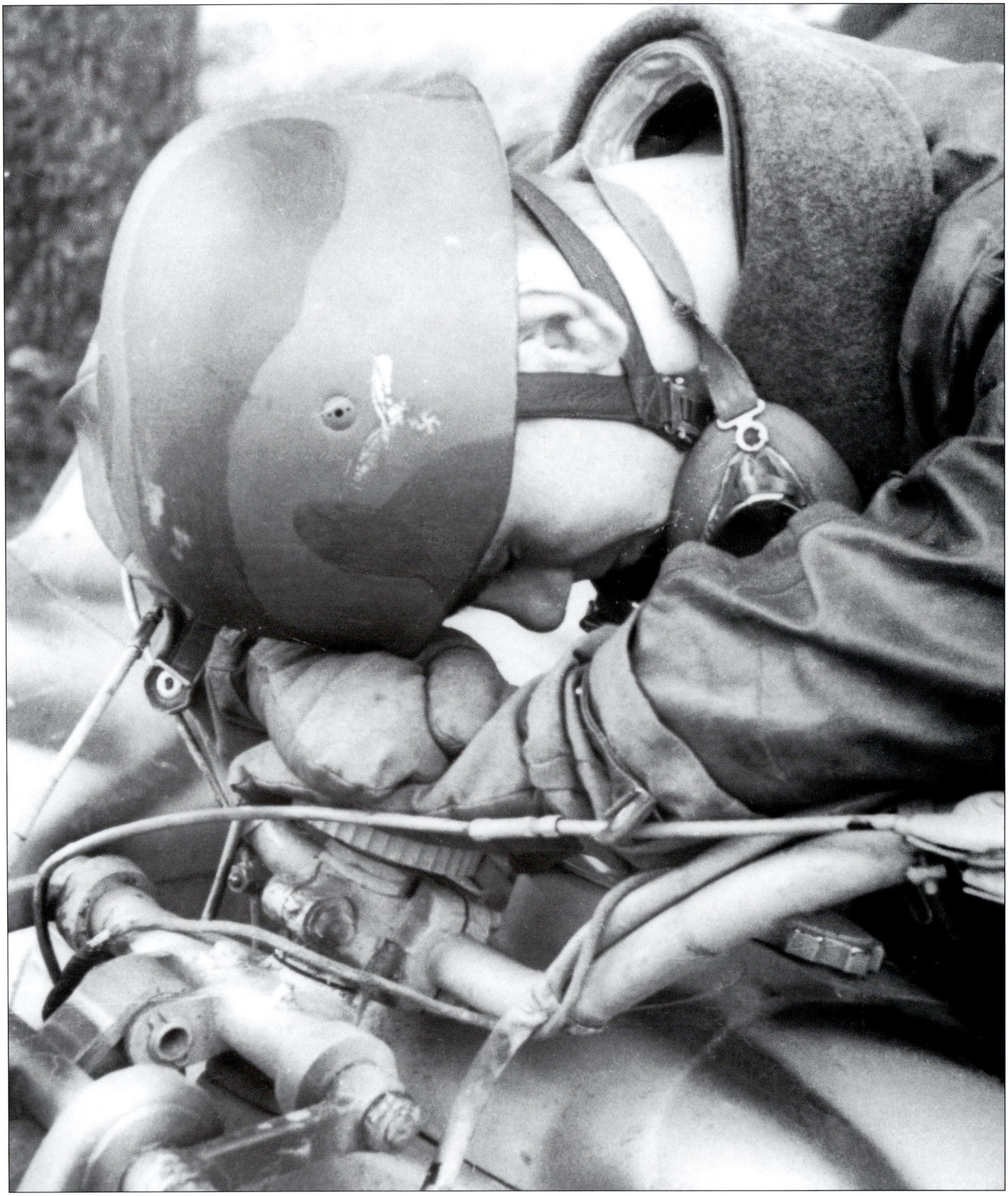

A tired *Fallschirmjäger* motorcyclist wearing an M38 with the regulation wartime matte grey finish, over which he has added patches of a solid colour, probably green. *(BA)*

Top: Men pictured in the ruins of Monte Cassino. The man on the left is wearing the drawstring cloth cover, while the man on the right has a much-used M38 over-painted in tan. *(Author)*

Bottom: Fallen *Fallschirmjäger* in the USSR prior to burial. It is interesting to note the white painted M38s, but also the fact that their helmets are to be buried with them. *(BA)*

Reinforcements rushing down the road from Rome to Anzio. *Fallschirmjäger* are wearing various tan-painted helmets. *(BA)*

White painted M38 helmets adorn the graves of *Sturmregiment Fallschirmjäger* killed in the USSR in January-February 1942. *(BA)*

A dozing Waffen-SS man wearing an M40 helmet that seems to be overpainted in tan with daubed patches of a darker colour, possibly green or brown. (Author)

Chapter Two: Textured Camouflage

■ M35, over-painted in light and dark green and brown over a sand/grit base. Note that a chicken wire cage was once attached. (Collection of David Asselin)

M35, originally painted dark grey-green, re-painted dark grey-green over a sand/grit base. The Wehrmacht eagle insignia has been left visible. (Collection of John Burnett)

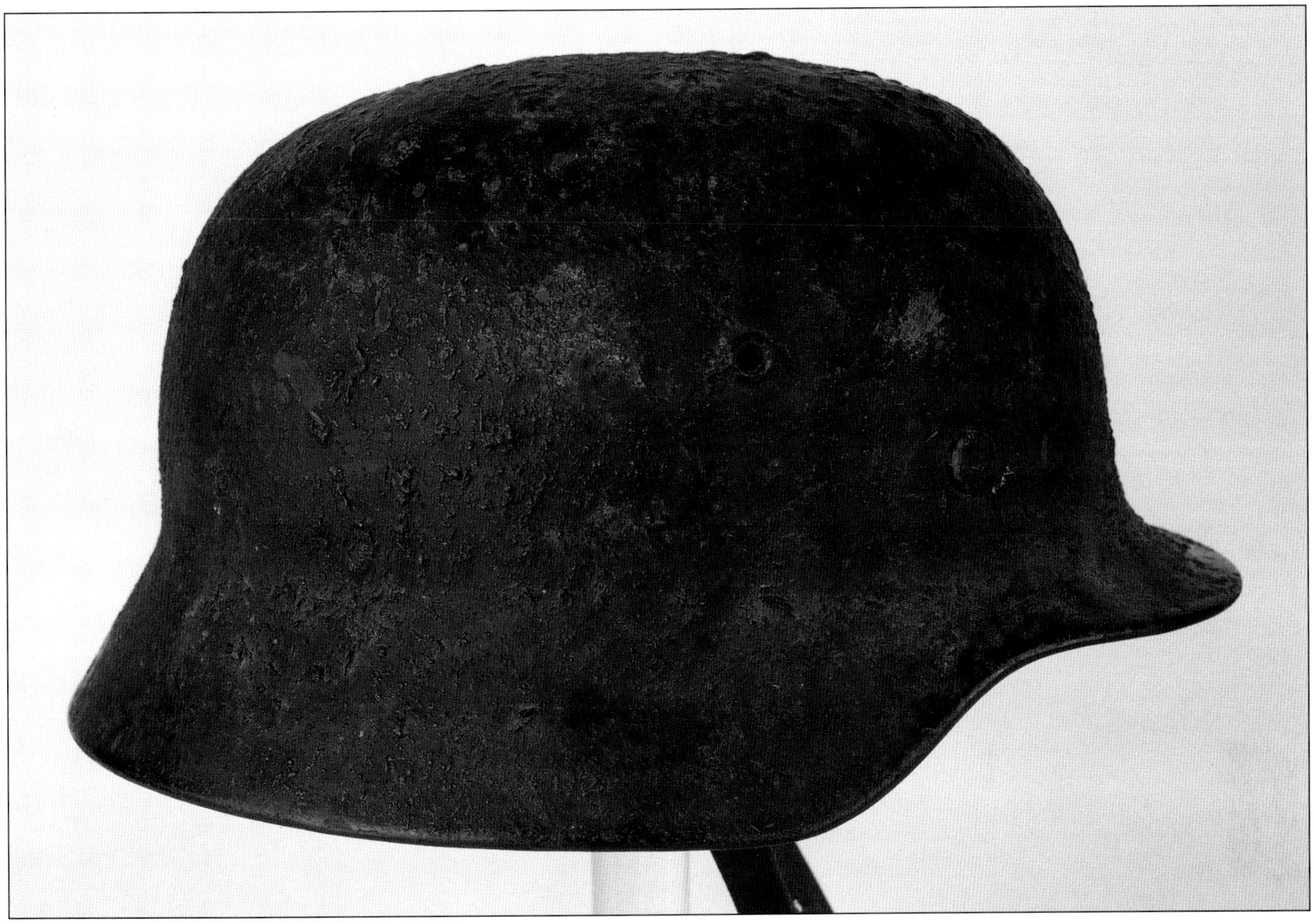

M35, originally painted pre-war apple green, completely overpainted in a light grey-green sand/grit mixture. It carries an Italian leather chinstrap and was painted with swastikas by the veteran who captured it. (Collection of Barry Smith)

M35, originally painted dark grey-green, re-painted dark grey-green over a sand/grit base. The Wehrmacht eagle insignia has been left visible. (Collection of the Imperial War Museum, London)

M40, originally painted dark grey-green, re-painted with geometric shapes of brown, green and tan in a sand/grit mix. (Collection of John Burnett)

M40, originally painted dark grey-green, re-painted light tan and over that brown and green in a sand/grit mixture. (Collection of David Asselin)

M40, originally painted dark blue-grey, re-painted light green and brown in a sand/grit mixture. (Collection of John Burnett)

M40, originally painted dark grey-green, completely overpainted in tan with a woodchip/grit mixture. Note the wear around the base of ther skull, suggesting that a band or strap was worn. (Collection of Brian Maederer - Military Collectibles)

M40, originally painted dark grey-green, re-painted large patches of brown in a sand/grit mixture. (Collection of John Burnett)

M42, originally painted dark grey-green, completely repainted green in a sand/grit mix with patches of tan. (Collection of John Burnett)

M42, originally painted dark grey-green, completely re-painted tan in a sand/grit mix with patches of brown and green. The Kriegsmarine eagle insignia is visible through the paint (Collection of Barry Smith)

M42, originally painted dark grey-green, completely over-painted in dark green with patches of tan in a heavy sand/grit mix. (Collection Brian Maederer - Military Collectibles)

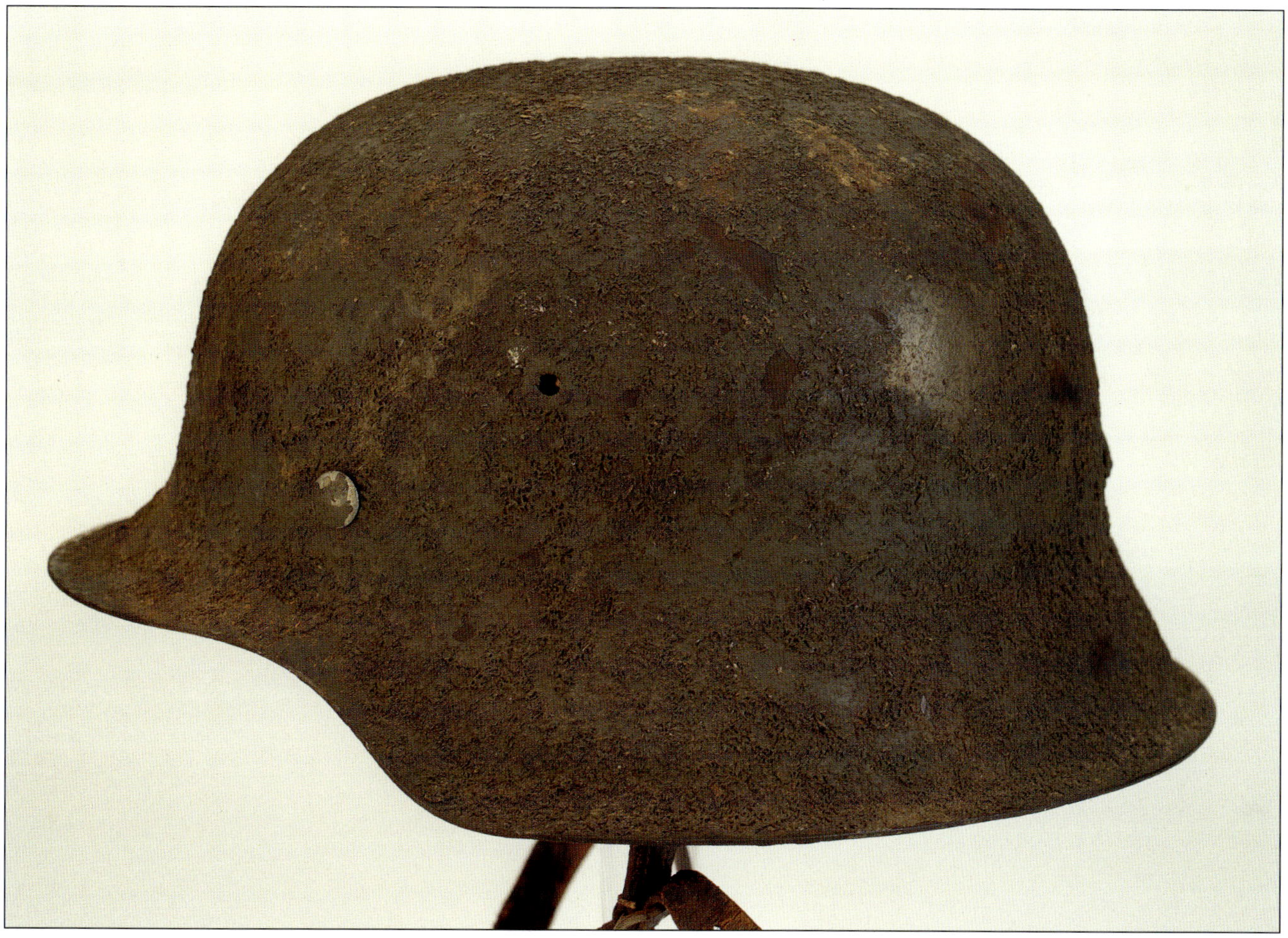

M42, originally painted dark grey-green, completely over-painted in tan with patches of brown and green in a heavy sand/grit mix. (Collection Brian Maederer - Military Collectibles)

M42, originally painted dark grey-green, completely over-painted in tan with a sand/grit mix and patches of thick brown and green paint. (Collection of John Burnett)

M42, originally painted dark grey-green, completely over-painted in tan with a sand/grit mix and patches of khaki. (Collection of Barry Smith)

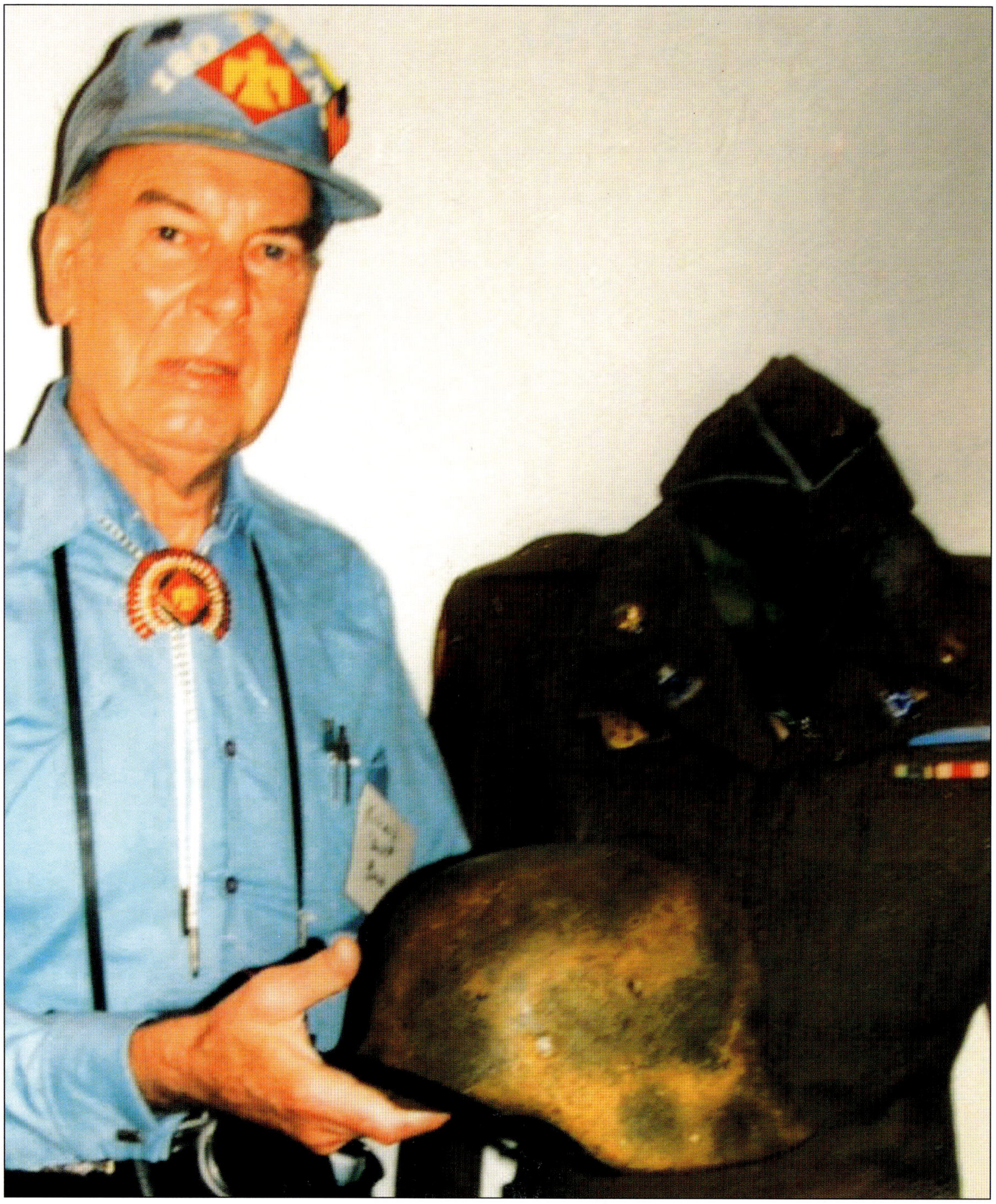

Veteran Technical Corporal Richard R. Bond, 180th Infantry regiment, 45th Infantry Division holding the helmet shown previously, which he took as a souvenir from a German MG42 position on hill 470, near Cassino, Italy. The period photograph shown opposite shows Bond's comrade Andy Davis in the MG position next to the same German helmet.

M42, originally painted dark grey-green, completely over-painted with a grey-green sand/grit mixture and over-painted again in tan. (Collection of Chris Jones)

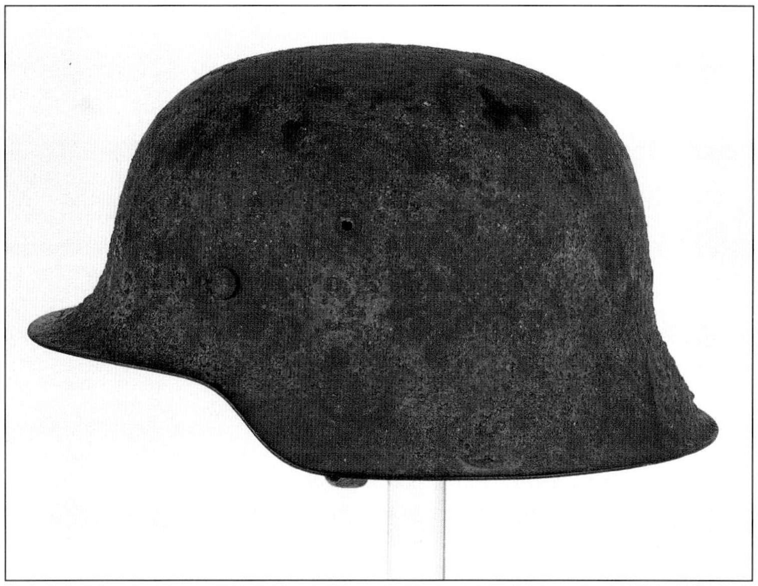

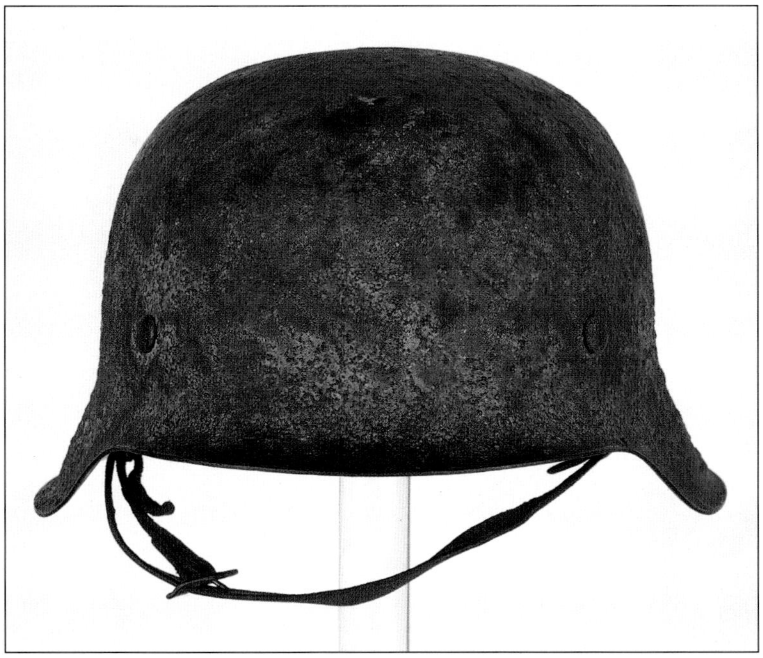

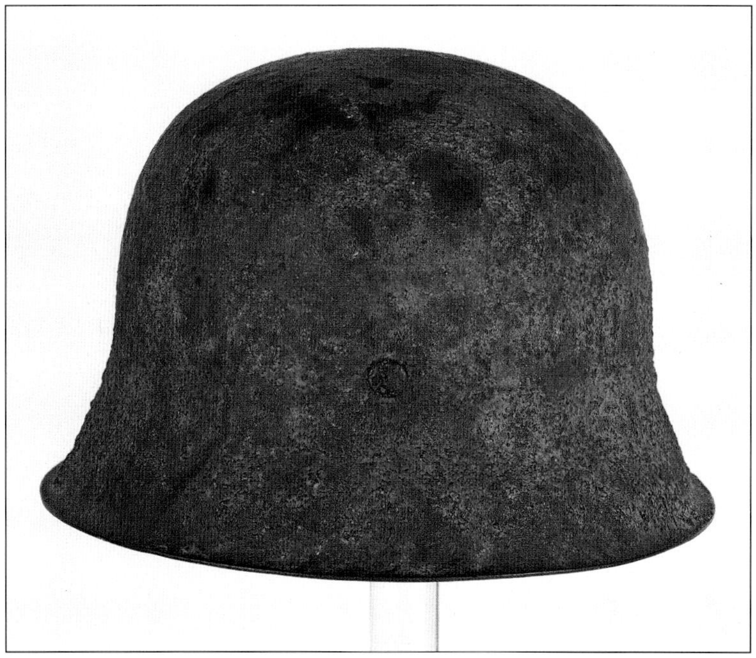

M38, probably originally painted pre-war apple green, completely over-painted in green with a heavy sand/grit mixture and patches of red-brown. The remains of wire clips suggest either wire or netting was attached at some time. (Collection of Chris Jones)

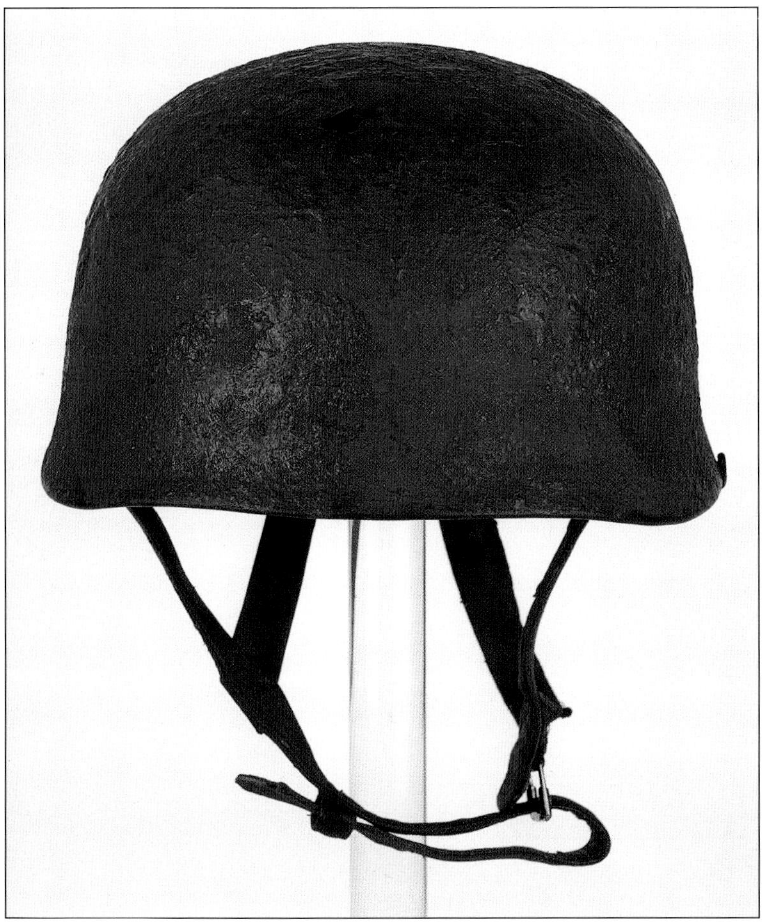

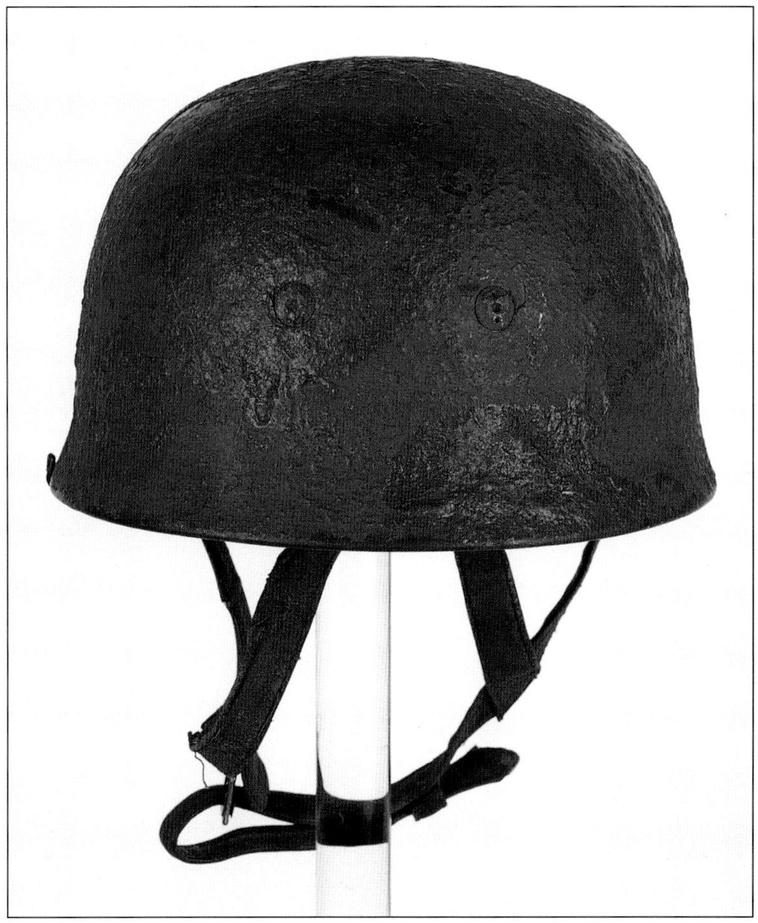

M38, originally painted dark grey-green, over-painted with green, tan and brown heavy sand/grit mixtures. (Collection of Chris Jones)

M38, originally painted dark grey-green, over-painted in light khaki then over-painted again in tan sand/grit mixture. Note the use of a standard M34/40/42 liner pin to replace a lost bolt. (Collection of the Imperial War Museum, London)

M38, originally painted dark grey-green, repainted with a tan sand/grit mixture and green stripes in a tortoise shell pattern. The Luftwaffe second pattern eagle insignia has been left visible. (Collection of the author)

M38, originally painted dark grey-green, entirely repainted in dark green heavy sand/grit mixture with patches of tan. The Luftwaffe second pattern eagle insignia has been left visible and the liner/ventilation bolts are the mid-to-late war vented slotted type. The late-war chinstrap is made from un-matched leather. (Collection of the author)

55
66

Period Photos

An army NCO wearing an M35 helmet. It has been covered in a mixture of grey-green matte paint and earth or sawdust to achieve a rough texture. *(BA)*

Top: This sniper group, photographed in September 1942, are all wearing well-used textured helmets. *(Author)*

Bottom: *Sturmregiment Fallschirmjäger* troops gather for a smoke in Tunisia. We see a variety of camouflages here: the early clip-on cloth cover; heavy texture finish; and texture with over-painted darker vertical stripes. (Note their tan-painted leather pouches and map-case.) *(BA)*

Another photograph from the same series in Tunisia. (This time note the tropical boots they are wearing.) *(BA)*

Top: A close-up, showing the heavy texture finish (paint mixed with sand) over-painted with darker patches. (Note the dark hand-painted shooting star insignia of the *Sturmregiment* over the eagle insignia of the man on the right.) *(BA)*

Bottom: Again, a nice clear shot of our *Sturmregiment* men. *(BA)*

A detail of heavy textured paint and sand mixture applied to an M38 helmet. *(BA)*

NOTES

NOTES

Also from the Publisher

THE COMPANION VOLUME

German Camouflaged Helmets of the Second World War

Branislav Radovic

Volume 2:

Wire, Netting, Covers, Straps, Interiors, Miscellaneous

Size: 9"x12", over 550 color and b/w photographs

328 pages, hard cover

ISBN: 0-7643-2106-4 $79.95